Brain
Puzzler's
Delight

· · · · · · · · · · · · · · · · ·

E.R. Emmet

Sterling Publishing Co., Inc. New York

To Richard and Brian
and the many others
on whom these were tried out

Library of Congress Cataloging-in-Publication Data

Emmet, E. R. (Eric Revell)
　[101 brain puzzlers]
　Brain puzzler's delight / E.R. Emmet.
　　p.　　cm.
　"This edition is adapted from the book of the same title originally
published by Emerson Books"—T.p. verso.
　ISBN 0-8069-8816-9
　1. Mathematical recreations.　I. Title.
　[QA95.E4　1993]
　793.7'4—dc20　　　　　　　　　　　　　　　　　　　　　93-8555
　　　　　　　　　　　　　　　　　　　　　　　　　　　　　　　　CIP

10　9　8　7　6　5　4　3　2　1

Published 1993 by Sterling Publishing Company, Inc.
387 Park Avenue South, New York, N.Y. 10016
This edition is adapted from the book of the
same title originally published by Emerson Books
© 1970 by E.R. Emmet. This edition
© 1993 by Sterling Publishing Company, Inc.
Distributed in Canada by Sterling Publishing
% Canadian Manda Group, P.O. Box 920, Station U
Toronto, Ontario, Canada M8Z 5P9
Distributed in Great Britain and Europe by Cassell PLC
Villiers House, 41/47 Strand, London WC2N 5JE, England
Distributed in Australia by Capricorn Link Ltd.
P.O. Box 665, Lane Cove, NSW 2066
Manufactured in the United States of America
All rights reserved

Sterling ISBN 0-8069-8816-9

◆ Contents

Preface 5

1. Warming Up 7

Predictions • The Architect's Shirt • Who Killed
Popoff? • River Road • Schoolmates • Vests and
Vocations • The Poison Spreads

2. Missing Digits 13

Hitler and Goering • Long Division #1 • Long
Division #2 • Long Division #3 • Long Division #4
• Long Division #5 • Long Division #6 • Add and
Subtract #1 • Add and Subtract #2 • Long Division
#7 • Addition and Division

3. Not So Easy! 23

The Christmas Compensation Club • The Years
Roll By • Sinister Street • Youthful Ambitions •
Uncle Knows Best • Out the Window and Over the
Wall • The Willahs and the Wallahs • Tom, Dick and
Harry • Monogamy Comes to the Island

4. Challenges 33

Uncles and Cousins • Intellectual Awareness • An
Intelligence Test • Birthdays • Easter Parade • Trips
Abroad • Clubs and Careers • The Five Discs •
Gowns for the Gala • Salamanca Street • Around
the Bend • Island of Indecision • Country Crescent

Solutions 47

Index 96

◆ Preface

The making up of puzzles has become for me over the years a sort of compulsive habit. Some of the problems in this book obviously date, but most of them have been put together recently. I am grateful to the *Sunday Times* for permission to reprint "Trips Abroad," which appeared in their "Brain-Teasers." None of the others appeared in print before this book.

People differ considerably in their attitude to and aptitude for problems of this kind. Some find them an engaging challenge; for others the mind automatically switches off. I believe that most people can get pleasure and satisfaction from such intellectual exercises if they can learn to tackle them with some measure of confidence. The puzzles in this book are of varying degrees of difficulty, and I hope that many readers may be encouraged, by a satisfactory solution of some of the easier ones, to tackle the more complicated varieties. In no case is any mathematical knowledge required beyond the most elementary. The fully worked-out solutions are of course designed to help and encourage those who have little experience of this kind of thinking.

Different readers will be interested in different kinds of problems. I have tried to make each puzzle independent, so that neither the question nor the solution depends in any significant way on anything that has gone before.

I am grateful to a large number of people, mostly members of my mathematical sets at Winchester, on whom the problems have been tried out. But most especially I owe a great debt of gratitude to Richard Longmore and Brian Orange, who have each checked all the problems and solutions in detail, have discovered a number of errors, and have made a great many most helpful suggestions for clarification. I need hardly say that any errors that may still remain are entirely my responsibility.

The production of the manuscript for a book like this is inevitably complicated and difficult, and I am most grateful to Mrs. J.H. Preston and Miss P. Kerswell for the skill and patience with which they have done this job.

E.R. Emmet

1. Warming Up

◆ Predictions

George, John, Arthur and David are married, but not necessarily respectively, to Christine, Eve, Dana and Rose. They remember that at a party years ago various predictions had been made. George had said that John would not marry Christine. John had said that Arthur would marry Dana. Arthur predicted that whoever David married, it would not be Eve. David, who at that time was more interested in horses than matrimony, had predicted that Gallant Dancer would win the Kentucky Derby. The only one to predict correctly was the man who later married Dana.

Who married whom? Did Gallant Dancer win the race?

Answer on page 48.

◆ The Architect's Shirt

Jones, Brown, Smith and Rodriguez were, not necessarily respectively, an architect, a photographer, a lawyer and a plumber. They wore, again not necessarily respectively, red, blue, black and green shirts.

The architect beat Brown at billiards. Smith and the lawyer often played golf with the men in black and green shirts. Jones and the plumber both disliked the man in the green shirt, but this was not the photographer, as he wore a red shirt.

What color was the architect's shirt? And what were the occupations and colors of the shirts of the other men?

Answer on pages 48–49.

◆ Who Killed Popoff?

Allan, Brenda and Charles had so often expressed their opinion of Professor Popoff that when he was found murdered (stabbed to death with a dagger, but in a very refined way) it was natural that they would be suspected. In fact, for reasons that we don't need to go into now, it is certain that one of them is guilty. They made statements as follows:

ALLAN: 1. I hadn't seen Popoff or had any contact with him for a week before his unfortunate demise.
2. Everything that Brenda says is true.
3. Everything that Charles says is true.

BRENDA: 1. I have never handled a dagger.
2. Everything that Allan says is false.
3. Everything that Charles says is false.

CHARLES: 1. Allan was talking with Popoff just before he was killed.
2. Brenda has handled a dagger.
3. I have for a long time respected Popoff more than most people realize.

Looking back on the tragic event now, it is interesting to see that Allan and Brenda both made the same number of true statements. (This number can be anything from 0 to 3.)

Who killed Popoff?

Answer on pages 49–50.

◆ River Road

Andy, Brad, Cole and Doug all live in different houses on River Road.

By a curious coincidence, the age of each man is either seven greater or seven less than the number of his house. All of them are over 15 years old and less than 90: their ages are all different.

Andy said that the number of Brad's house was even and Brad remarked that the number of his house was greater than that of Doug's. "My age," he added proudly, "is a perfect cube."

Cole said that the number of his house was greater by 3 than that of Andy, and that Doug's age was an exact multiple of Andy's age.

Doug, who has an unfortunate habit of complicating things, said that Brad's age was either 27 or an even number other than 64. "Furthermore," he commented, "Cole does not live at number 19."

Unfortunately, these remarks were not all true. It was interesting to note that remarks made by those who lived in even-numbered houses were false, and remarks made by those who lived in odd-numbered houses were true.

What are their ages and the numbers of their houses?

Answer on pages 50–51.

◆ Schoolmates

Alice, Brett, Catherine and Deirdre went to school together. They became, but not necessarily respectively, an author, a biologist, a cartoonist and a doctor. Years before, they belonged to A, B, C and D sororities, and they came from Australia, Brazil, Canada and Denmark. The letters of each woman's house, the initial letters of her profession, her home and her name are all different from each other. The doctor had never been to Brazil, and the biologist had never been to Canada.

Back at school, Catherine, the girl from Australia and the biologist used to spend all their spare time together.

What was the profession, the home and the house of each of them?

Answer on pages 51–52.

◆ Vests and Vocations

Ms. Baker, Ms. Carpenter, Ms. Hunter and Ms. Walker were, not respectively, a baker, a carpenter, a hunter and a walker by profession. They each owned a vest—again not respectively, one brown, one crimson, one heliotrope and one white. No woman's profession was the same as her name, and the color of the vest each woman owned began with a letter that was different from the initial letter of both her name and her profession.

Ms. Hunter and the professional walker had lunch together regularly. Ms. Hunter, rather curiously, violently disliked the color brown and would never wear a brown vest. Ms. Carpenter was the baker.

Find the profession of each woman and the type of vest she owned.

Answer on pages 52–53.

◆ The Poison Spreads

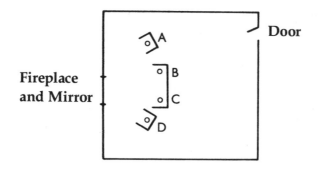

Colonel Chutney has just been found dead in the dining room of his club, poison in his wine.

Four men, seated as above on a sofa and two armchairs around the fireplace in the lounge, are discussing the murder. Their names are Scott, Bowen, Johnson and Robinson. They are, not necessarily respectively, a general, a headmaster, an admiral and a surgeon.

1. The waiter has just poured out a glass of port for Johnson and a cup of tea for Bowen.

2. The general looks up into the mirror over the fireplace. He sees the door close behind the waiter. He then turns to Robinson, who is next to him, and starts talking.

3. Neither Scott nor Bowen has any sisters.

4. The headmaster is a teetotaller.

5. Scott, who is sitting in one of the armchairs, is the admiral's brother-in-law. The headmaster is next to him on his left.

6. Suddenly, a hand stealthily puts something in Johnson's glass. It is the murderer again. No one has left his seat. Nobody else is in the room.

Who is the murderer? What is the profession of each man, and where is he sitting?

Answer on pages 53–54.

2. Missing Digits

◆ Hitler and Goering

Add Hitler to Goering and get—what? Letters substitute for digits in the sum below. The same letter stands for the same digit wherever it appears, and different letters stand for different digits.

```
    H  I  T  L  E  R
    G  O  E  R  I  N  G
  ──────────────────────
 H  T  T  L  L  H  H  H
```

Find the digits the letters represent.

Answer on pages 54–55.

◆ Long Division #1

Fill in the missing digits, including the figures in the quotient, in the following division sum:

```
          _ _ _ _ _ 5 _ _ _ 8
  _ _ _ ) ─────────────────────
          _ _ _
          ─────
            _ _ _ _
            _ _ _
            ─────
              _ _ _ _
              _ _ _ _
              ───────
                _ _ _ _
                _ _ _ _
```

Answer on pages 55–56.

◆ Long Division #2

represents a decimal point.

. __ . indicates that these three decimal figures recur.

Find the missing digits.

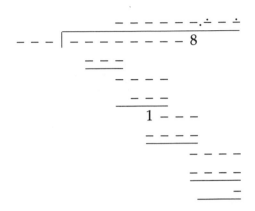

Answer on pages 56–57.

◆ Long Division #3

Another division sum. Note that the two *a*'s stand for the same digit, but that doesn't mean that the same digit doesn't appear elsewhere.

Find the missing digits.

```
              _ _ _ _ _
        ┌─────────────
  _ _ │  _ _ _ _ _
        _ _
        a  a
        _ _
        _ 3 _
        _ _ _
              _ _
              _ _
              _ _
              _ _
```

Answer on pages 58–59.

♦ Long Division #4

In this division sum, *a* and *b* stand for digits that are such that $a - b = 2$. (These are not necessarily the only places where these digits occur.)

Find the missing digits.

```
      ┌─────────────
_ _   │ _ _ 5 _ _
        _ _
        _ _ _

        _ _ _
        ─────
          _ b _
          _ _ _
          ─────
            a _ _
            _ _ _
            ─────
```

Answer on pages 59–60.

◆ Long Division #5

In this division sum, *a* stands for the same digit wherever it occurs. But it does not necessarily follow that this digit does not occur elsewhere.

Find the missing digits.

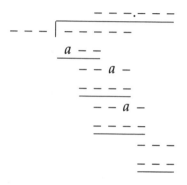

Answer on pages 61–62.

◆ Long Division #6

There are two alternative solutions to the division sum below.

Find both solutions.

```
        _ . _ _
 _ _ | _ _ _
       _ _
      _____
       _ _
      _ _
      _ _
       _ _
      _ _
```

Answer on pages 62–63.

◆ Add and Subtract #1

(i) and (ii) represent the same two numbers. In one case, they are added together, and in the other case they are subtracted from each other. No 0s appear anywhere. The digits represented by the two a's are the same. ($b - c = 1$).

In neither (i) nor (ii) does the same digit appear more than once.

Find all the digits.

(i)
```
  _ _ c
  _ b
  _____
  _ a
```

(ii)
```
  _ _ _
  _ _
  _____
  _ _ a
```

Answer on pages 63–64.

♦ Add and Subtract #2

(i) and (ii) represent the same two numbers being added in one case and subtracted in the other. In (i), all the digits are different. In (ii) there are two pairs of digits, which are the same, represented by the two a's and the two b's, but otherwise the digits are all different. The digits in the result of the addition are all different from any of the digits in the result of the subtraction.

Find the missing digits.

(i) $-\ -\ -$ (ii) $-\ -\ -$

$-\ -$ $a\ \ b$
$\overline{}$ $\overline{}$
$-\ -\ -\ -$ $b\ -\ a$

Answer on pages 64–65.

◆ Long Division #7

In this division sum, the three *a*'s stand for the same digit.
$b - c = 2$; *a*, *b*, *c* and *d* are all different digits. (The places indicated are not necessarily the only places where *a*, *b*, *c* and *d* occur.)

Find the missing digits.

```
              a _ _ _ _ _ _._ _
    _ _ | _ _ _ _ d _ _ _ a
          _ _
          b c _
          _ _ _
            _ _
            _ _
            a _ _
            _ _ _
              _ _
              _ _
              _ _
              _ _
                _ _ _
                _ _ _
                _ _ _
```

Answer on pages 65–66.

◆ Addition and Division

(i) shows a four-figure number and a two-figure number added together.

(ii) shows the same four-figure number multiplied by 100, divided by the same two-figure number. (Note that the division leaves a remainder.)

Find the missing digits.

(i)   – – – –   (ii) – – ⟌ – – – – – 0 0

        – –   – –
      ‾‾‾‾‾   ‾‾‾
    – – – – –   – – –

                          – –
                        ‾‾‾‾
                          – – –

                          – – –
                        ‾‾‾‾
                            –

Answer on pages 66–68.

3. Not So Easy!

◆ The Christmas Compensation Club

Gina, Hal, Irene, John and Kate are members of the Christmas Compensation Club—open only to those who suffer from having their birthday on Christmas Day and have not yet reached the age of 90.

Last Christmas, Kate was older than Irene by three times as much as she was older than Hal, and John was ten percent younger than Hal and 20 percent older than Irene. The difference between the ages of Gina and Kate is the same as the difference between the ages of John and Irene (and in the same sense).

Find their ages.

Answer on pages 68–69.

◆ The Years Roll By

Annie, Bill, Chris, Dirk and Erika, having nothing better to do, are making remarks about their ages—as follows:

ANNIE: Erika is 27.
  BILL: I'm 81.
CHRIS: Bill is 61.
  DIRK: (i) Annie is 57.
        (ii) What Chris says is false.
ERIKA: (i) Bill is older than Annie.
        (ii) Dirk is 30 years younger than Chris.

None of them is less than 10 or more than 99.

Remarks made by anyone who is 50 or over are true, unless the person's age is a perfect square. Remarks made by anyone who is under 50 are false, unless the person's age is a perfect cube.

Find all their ages.

Answer on page 69.

◆ Sinister Street

Agnes and Beatrice live in different dwellings in Sinister Street, which has houses numbered from 1 to 99. Neither of them knows the number of the other's house. Xerxes lives on the same street—in a different house—and the ladies are anxious to know where.

Agnes asks Xerxes two questions:

1. Is your number a perfect square?
2. Is it greater than 50?

Having heard the answers, she claims that she knows Xerxes' number and writes it down. But she is wrong, which is not surprising, considering that only the second of Xerxes' answers is true.

Beatrice, who has heard none of this, then asks Xerxes two different questions:

1. Is your number a perfect cube?
2. Is it greater than 25?

And, like Agnes, when she has heard the answers, she claims that she knows where Xerxes lives. But again, like Agnes, she is wrong, for Xerxes has only answered the second question truthfully.

If given the additional information that Xerxes' number is less than that of Agnes or Beatrice, and that the sum of their three numbers is a perfect square multiplied by two, you should be able to *discover where all three live.*

Answer on pages 69–70.

◆ Youthful Ambitions

Adam, Biff, Chad, Dan and Earl are married, but not respectively, to Alison, Barbara, Chita, Danielle and Erin. The birthplaces of the five men were, not respectively, Amherst, Bristol, Chicago, Delhi and Edmonton; and, a most intimate and revealing detail, their youthful ambitions were, again not respectively, to become an astronaut, a banker, a chemist, a dog breeder and an Egyptologist.

For each man, his name, the name of his wife, his home town and his youthful ambition all begin with different letters.

The five ladies make remarks (which, unfortunately, are not all true) as follows:

> ALISON: The would-be dog breeder is married to Barbara.
>
> BARBARA: 1. The man who wanted to become an Egyptologist is not Danielle's husband.
> 2. The man who was born in Bristol has always wanted to be a dog breeder.
>
> CHITA: The man who was born in Amherst is married to Barbara.
>
> DANIELLE: The would-be banker is not my husband.
>
> ERIN: 1. The man who was born in Amherst is not Biff.
> 2. The man who was born in Chicago wanted to become an Egyptologist.

It is interesting to notice that, in these remarks, when the subject of the sentence is a man whose name begins with a letter that comes before the initial letter of the speaker in the alphabet, the sentence is *true*; if the initial letter of the subject's name comes *after* the initial letter of the speaker's name, the sentence is *false*. In no case is the initial letter of the subject's name the same as the initial letter of the speaker's name.

Find, for each man, the name of his wife, his hometown and his youthful ambition.

Answer on pages 70–72.

◆ Uncle Knows Best

Uncle Chester is a bit hard of hearing, so that when his nephew Bernard asked various neighbors who live in separate houses in Country Crescent some questions about the numbers of their houses, he failed to hear the answers—though he heard the questions, all right. Bernard has lived in Country Crescent for some time, and his uncle knows his number, but Chester has only just bought a vacant house there and Bernard doesn't know about it.

The Crescent has houses numbered from 5 to 105.

Bernard asked the same three questions of three people who live in separate houses in the Crescent—X, Y and Z:

1. Is the number of your house a multiple of 4?
2. Is it a perfect square?
3. Is it a multiple of 9?

No two sets of three answers are the same in every respect.

After hearing X's answer, Bernard says to him, "If I knew whether the number of your house was greater than 83, I would know what it is."

(Chester hears his nephew say this and is able to write down X's number correctly.)

After hearing Y's answer, Bernard says to him, "If I knew whether your number was greater than 50, I could tell you what it is."

(Chester hears his nephew say this, and as he happens to know that Y's number is greater than his own, he is able to write it down correctly.)

After hearing Z's answer, Bernard says to him, "If I knew whether the number of your house was greater than 30, I could tell you what it is."

(Chester hears this and notes with interest that his own number is certainly less than Z's. But he too has no way of deciding whether Z's number is greater than 30. However, he is anxious to keep the reputation he is building up for intelligence and intuition. So, he guesses that it is greater than 30 and writes it down. Fortunately, he is quite right.)

What are the numbers of the houses of Bernard, Chester, X, Y and Z?

Answer on pages 73–74.

◆ Out the Window and Over the Wall

Alex, Baxter and Clay are in prison. It doesn't matter why. What is important is that they get out as soon as possible. They have made arrangements for getting through the window of their cell and over the outer wall at X. But they can only hope to escape unseen if they do it in the dark, one at a time, and if each man has a clear two minutes with all the warder sentries at least 100 yards away.

They reckon that it is dark enough for their purposes at 9 p.m., and at that time a new group of sentries comes on duty. There are two sentries inside the outer wall. One starts at R1, 300 yards west of a point outside their window, marches to R2, exactly outside their window, and turns

Plan of possible exit

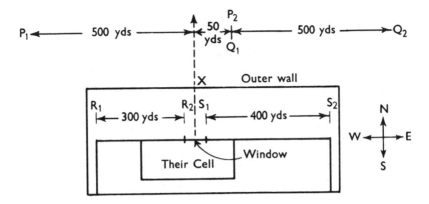

around. Another starts at S1, outside their window, marches to S2, 400 yards to the east, and turns around.

There are also two sentries outside the outer wall. One starts at P1, 500 yards west of a point outside their window, marches 550 yards to P2, 50 yards east of a point outside their window, and turns around. The other sentry starts at Q1 (the same point as P2), marches 500 yards east to Q2, and turns around.

The sentries start at 9 p.m. from the points P1, Q1, R1, S1 and march up and down their beats at a regular speed of 100 yards per minute. (Time spent turning around can be ignored.)

At what times should Alex, Baxter and Clay attempt to escape?

Answer on page 74.

♦ The Willahs and the Wallahs

The inhabitants of a small island in the Pacific are divided into two tribes—the Willahs and the Wallahs. Every inhabitant has a different personal number. The numbers of the Willahs are all primes and the numbers of the Wallahs are not primes.

Remarks made to people belonging to the same tribe as the speaker are always true; remarks made to someone belonging to a different tribe from the speaker are never true.

A, B, C, D, E and F are six natives, three belonging to each tribe. They make remarks as follows:

A to B: D is a Wallah.
B to D: F's number is halfway between yours and C's.
C to E: I'm a Willah.
C to F: D's number is halfway between yours and A's.
D to A: Our numbers are both less than 50.
E to A: I'm a Wallah.
E to B: My number is 35.
F to C: My number is greater than yours by 10.

Find to which tribe each person belongs and the numbers of as many of them as possible.

Answer on page 75.

◆ Tom, Dick and Harry

There are three tribes on the Island of Imperfection—the Pukkas, who always tell the truth; the Wotta-Woppas, who never tell the truth; and the Shilli-Shallas, who make statements that are alternately true and false or false and true.

An explorer lands on the island and questions three natives—Tom, Dick and Harry—as follows:

He asks Tom: Which tribe do you belong to?
Tom answers: I'm a Pukka.
He asks Dick: (i) Which tribe do you belong to?
Dick answers: I'm a Wotta-Woppa.
He asks Dick: (ii) Was Tom telling the truth?
Dick answers: Yes.
He asks Harry: (i) Which tribe do you belong to?
Harry answers: I'm a Pukka.
He asks Harry: (ii) Which tribe does Tom belong to?
Harry answers: He's a Shilli-Shalla.

To which tribe does each man belong?

Answer on page 76.

◆ Monogamy Comes to the Island

All the inhabitants of the island are still members of one of three tribes: the Pukkas, who always tell the truth; the Shilli-Shallas, who make statements that are alternately true and false (or false and true); and the Wotta-Woppas, who never tell the truth.

To prevent the unsatisfactory results of inbreeding, it has recently been decreed that there should be no marriages between people of the same tribe. It has also been laid down that in future there shall be one man/one wife.

The ladies have cheered up considerably now that they are to have a husband each. The four with whom our story deals are called Eager, Friendly, Glorious and Happy. The names of their husbands, not necessarily respectively, are Sordid, Tired, Ugly and Vulgar.

The men have not been talking much recently, and the ladies speak as follows:

EAGER: 1. Sordid is married to Glorious.
          2. Sordid is a Wotta-Woppa.
          3. Vulgar is not a Wotta-Woppa.
FRIENDLY: 1. Glorious is not a Pukka.
          2. My husband is a Pukka.
          3. My husband is Tired.
GLORIOUS: 1. Happy is married to a Wotta-Woppa.
          2. Ugly is a Pukka.
          3. Eager is a Wotta-Woppa.
HAPPY: 1. Friendly is a Pukka.
          2. Glorious is a Wotta-Woppa.
          3. Eager is a Shilli-Shalla.

Find out who is married to whom, and the tribes to which they all belong.

Answer on pages 76–77.

4. Challenges

♦ Uncles and Cousins

They always talked about themselves in the Company as one big happy family, and that is in fact literally true. Alfred, Barry, Carl, Donald, Edward, Ferdie and Greg can all be represented on one family tree so that every male who appears on the tree is one of the seven. Their jobs are, not necessarily respectively, Door-Shutter, Door-Opener, Door-knob Polisher, Sweeper-Upper, Bottle-Washer, Chief Cook and Worker. Except for Carl, who remains silent, they all make remarks that are true, as follows:

ALFRED: My brother-in-law is the Chief Cook.

BARRY: 1. My father's brother is Greg's sister's father-in-law.
2. My brother is the Door-Opener.

DONALD: 1. My father is the Bottle-Washer.
2. Edward's son is the Door-Shutter.

EDWARD: 1. Donald's brother is the Doorknob-Polisher.
2. My father is the Worker.

FERDIE: My nephew is the Door-Shutter.

GREG: Barry is the Door-Shutter's cousin.

Find their jobs and how they are all related to each other.

Answer on pages 78–79.

◆ Intellectual Awareness

Alfred, Barry, Carl, Donald and Edward have once more been indulging their insatiable urge for competition. This time they have been competing against each other for Intellectual Awareness—with the Managing Director of course as Judge. As a result, they finish in a certain order (no ties).

Alfred says: "I know Barry's place as well as my own—I was higher than he was—but no one else's. If I knew that Carl was three places higher than Donald—which is about what I would expect—I would know the places of all five of us."

Edward, who has been listening to Alfred's remarks, has been told no one's place except his own. But he feels quite sure that, as he puts it: "The half-witted and incompetent Donald could not possibly have been higher than Barry." (He was quite right—Donald was not higher than Barry.) After a pause for reflection, Edward says, "I can now write down all our places."

He does so, and he is quite right.

What was the order of finish?

Answer on pages 79–80.

◆ An Intelligence Test

My old friends Annette, Bonnie, Candice, Des and Ernie have just taken an intelligence test and scoring is complete. (There were no ties.)

They made various remarks—not all true, unfortunately—about the places of themselves and others. On examining these remarks, I noticed the curious fact that if the speaker had placed higher in the test than the person or persons mentioned, the remark was true. But if the speaker had placed lower than the person or persons mentioned, the remark was false. When two people were mentioned, in no case was the speaker between them.

As I do not intend to reveal to you which remarks were made by which person, I have called the speakers P, Q, R, S and T in no particular order. (Remarks that the speaker makes about himself are all made in the first person.)

P: I was higher than Bonnie and Candice.

Q: 1. Neither Annette nor Candice was fifth.

2. Ernie was second.

R: Candice was higher than Ernie.

S: Des was first.

T: 1. Ernie was fourth.

2. Candice was lower than Annette.

Find the identities of P, Q, R, S, T, and the order in which they finished.

Answer on pages 80–81.

♦ Birthdays

Andrea, Bridie, Constance, Darlene and Edwina have recently discovered the remarkable fact that all their birthdays are on the same day, though their ages are all different. On their mutual birthday (when their ages of course are all an exact number of years), they're having a typical conversation about it. Here are some of the things that I overheard.

Andrea said to Bridie: "Your age is exactly 70 percent greater than mine."

Bridie said to Constance: (i) "Edwina is younger than you." (ii) "The difference between your age and Darlene's is the same as the difference between Darlene's and Edwina's."

Constance said to Andrea: (i) "I'm ten years older than you." (ii) "Bridie is younger than Darlene."

Constance said to Darlene: "The difference between my age and yours is six years."

Darlene said to Bridie: "I'm nine years older than Edwina."

Edwina said to Bridie: "I'm seven years older than Andrea."

Knowing all their ages as I do, I realized at once that they were not all telling the truth. On analyzing their remarks, I discovered the interesting fact that when speaking to someone older than themselves, everything they said was true, but when speaking to someone younger, everything they said was false.

Find all their ages.

Answer on page 82.

◆ Easter Parade

Easter was a glorious Sunday, and the employees of the Company were going out for a stroll. Alfred, Barry, Carl, Donald, Edward, Ferdie and Greg had with them their wives, who are called, not respectively, Amy, Beth, Cleo, Deb, Enid, Fifi and Gay.

The wives are wearing, again not respectively, elegant hats that are florally decorated with Aspidistras, Begonias, Crocuses, Dahlias, Edelweiss, Fuchsias and Gentians.

Each woman's husband and flower have different initial letters that are not the same as the initial letter of the woman's name.

Each of the 14 people either always tells the truth or never does. Of each married couple, one member and one only is a liar.

The men are all speechless with embarrassment at their wives' hats, and the women do the talking, as follows:

AMY: 1. When asked whether Gay was wearing a Fuchsia hat, Greg said *No*.
2. When asked whether Edward's wife was wearing a Begonia hat, Donald said *Yes*.
3. When asked whether Beth was wearing a Fuchsia hat, Cleo said *Yes*.

BETH: 1. Alfred is not married to Enid.
2. When asked whether Ferdie was married to Gay, Cleo's husband said *Yes*.
3. When asked whether Greg's wife was wearing a Fuchsia hat, Ferdie said *Yes*.

CLEO: 1. Deb is wearing a Begonia hat.
2. My husband is Greg.
3. Enid's husband is a liar.

DEB: 1. When asked whether Donald was married to Cleo, Alfred said *Yes*.

2. Edward is a liar.

3. Ferdie's wife is wearing a Crocus hat.

ENID: 1. Alfred always tells the truth.

2. Carl's wife is not wearing an Edelweiss hat.

3. When asked whether she was Edward's wife, Cleo said *No*.

FIFI: 1. Donald's wife is wearing an Aspidistra hat.

2. When asked whether Deb was Barry's wife, my husband said *No*.

GAY: 1. When asked whether Fifi was wearing a Begonia hat, Barry said *Yes*.

2. Barry's wife is wearing an Aspidistra hat.

Find for each man his wife's name and the floral decoration on her hat, and find which of the 14 tell the truth and which tell lies.

Answer on pages 83–85.

◆ Trips Abroad

Anthony, Ben, Curt, Derek and Ellery went with their wives for vacations abroad last year. Bologna, Copenhagen and Dublin were the destinations of three of the pairs, but the other two ventured farther afield to Auckland and Ethiopia. I knew that the names of the five wives were Ava, Brigitte, Crystal, Darcy and Emily, but the only information I had about who was married to whom was that for each pair, the names of the husband, the wife and last year's vacation destination all began with different letters.

In an attempt to discover more, I talked with three of the ladies.

Brigitte told me that she was not married to Anthony and that she had heard from Ellery that Curt went to Dublin last year.

Darcy, however, firmly informed me that Curt went to Ethiopia and that Brigitte went to Dublin. "Unlike some people I could mention," she added darkly, but rather irrelevantly, "Anthony always tells the truth."

Crystal said that when her husband was asked whether Emily was married to Curt, he replied, "No." She went on to tell me that Derek went to Bologna.

When I had been told from another source the curious fact that of each of these married couples, one member always told the truth and the other never did, I was able to deduce the name of each man's wife and where they all went for their vacations.

Can you?

Answer on pages 86–87.

◆ Clubs and Careers

Arnold, Boris, Cyrus, Dudley and Edgar are members, not necessarily respectively, of five different clubs—the Boomers, the Amalgamated Brick Droppers, the Simple Life, the Better Life and the Longer Life. Their professions are, again, not necessarily respectively, Advertising Consultant, Sales Promoter, Purchasing Agent, Office Manager and Garbage Collector. In no case does anyone know the profession or the club of any of the others before our story starts.

And then, when they are all together, they begin to talk.

Arnold says that he is not a member of the Amalgamated Brick Droppers and that he is not a Sales Promoter.

Boris says that he does not belong to the Simple Life and is not an Office Manager.

Dudley says that he too is not a member of the Simple Life and also is not a Sales Promoter.

Cyrus has been listening to these remarks with his usual care and intelligence and says: "If I knew that the person who is a member of Simple Life was a Sales Promoter, I would know for certain that Arnold was a member either of Boomers or of Longer Life, and that Boris was either an Advertising Consultant or a Purchasing Agent."

Edgar has also been listening carefully and he is, if possible, even more intelligent than Cyrus. He says, "Dudley must be a member of either Boomers or of Longer Life, but I don't know which. And Dudley must be either the Office Manager or the Purchasing Agent."

With the additional information that the Purchasing Agent is not a member of the Simple Life, and that the Sales Promoter is not a member of the Boomers, you should be able to *find all their clubs and professions*.

Answer on pages 87–88.

◆ The Five Discs

(This is an adaptation of the famous "three disc" problem, but in a more complicated form.)

There are five men, A, B, C, D and E, each wearing a disc on his forehead selected from a total of five white, two red and two black. Each man can see the colors of the discs worn by the other four, but he is unable to see his own. They are all intelligent people, and they are asked to try to deduce the color of their own disc from the colors of the other four that they can see. In fact, they are all wearing white discs. After a pause for reflection, C, who is even more intelligent than the others, says, "I reckon I must be wearing a white disc."

By what process of reasoning could he have arrived at this conclusion?

Answer on page 89.

◆ Gowns for the Gala

Five young ladies, Priscilla, Quinta, Rachel, Sybil and Tess, were discussing what their gowns should be made of for the forthcoming ball.

Priscilla says: "I will wear lace unless Rachel wears muslin and Sybil does not wear nylon, in which case I will wear organdy, but not otherwise."

Quinta, who is rather bossy, says: "Rachel must wear nylon, unless Tess wears lace (in which case Rachel must wear organdy) or unless Tess wears muslin (in which case Rachel must wear muslin too)."

Rachel says: "I will wear lace unless Sybil wears muslin."

Sybil says: "If Priscilla wears lace, then I will wear organdy."

Tess says to Quinta: "If Priscilla does not wear lace, you will wear either lace or muslin—whichever of these two is worn less by the rest of us."

These remarks seem to be a happy blend of predictions, commands and expressions of intention. But, whichever they were, it is pleasant to record that they were all obeyed or proved to be true.

What did the young ladies wear at the ball?

Answer on pages 89–90.

◆ Salamanca Street

Goncle, Honcle and Izzie live in three different houses (between numbers 13 and 99 inclusive) in Salamanca Street. Jocko is an old friend who is staying with Goncle.

Goncle, Izzie and Jocko do not know the number of Honcle's house, but all four of them know the numbers of Goncle's and Izzie's (Izzie lives in number 49; the number of Goncle's house is odd, and is less than that of Izzie).

Goncle asks Honcle the following questions about the number of his house:

1. Is the number of your house bigger than mine?
2. Is it bigger than Izzie's?
3. Is it a perfect square?
4. Is it divisible by 3?

Izzie and Jocko hear these questions and the answers that Honcle gives.

Goncle thinks that Honcle's answers are all true. Whatever Izzie thinks is true in Honcle's answers, Jocko thinks is false, and vice versa. Jocko thinks that only Honcle's second answer is true. Honcle's answers are in fact alternately "Yes" and "No," but you are not told which comes first.

After a pause for thought, Goncle, Izzie and Jocko all say that they know the number of Honcle's house, but when asked what it is, their answers are all wrong.

Kipper, who has been listening, does not live in Salamanca Street. He is informed by Honcle that only two of his answers are true, and Honcle tells him which. Kipper, who already knows the numbers of Goncle's and Izzie's houses, is then able to announce the number of Honcle's house correctly.

What are the numbers of Goncle's and Honcle's houses?

What were the incorrect numbers of Honcle's house given by Goncle, Izzie and Jocko?

Answer on pages 90–91.

◆ Around the Bend

If you go around Cat's Corner in Country Crescent, you come to a group of new houses that are numbered from 101 to 200 inclusive. They are reserved for higher mathematicians. The three Fraction brothers—Proper, Improper and Vulgar—have recently moved into separate houses here, but they don't know where either of the others lives. They all take the view that mathematical information should be earned and not given away.

Proper, however, has persuaded Vulgar to write down the answers to three questions about the number of his house. They are:

1. Is your number a square?
2. Is it a cube?
3. Is it a multiple of 29?

Proper reads the answers and says, "If I knew that your number was greater than 150, I could tell you what it is."

Improper has read the questions and answers and has heard Proper's comment. From his observations, he has reason to believe that Vulgar's number is less than 150, and that the difference between the number of his house and that of Proper is less than 30.

After a little thought, he claims to be able to write down the numbers of the other two houses. He does so, but only Proper's number is correct, which is not surprising considering that only Vulgar's answer to question #2 was true, and that Improper's belief that Vulgar's number was less than 150 was incorrect.

What are the numbers of the houses of Proper, Improper and Vulgar?

Answer on pages 92–93.

◆ Island of Indecision

Each inhabitant of the Island of Indecision belongs to one of four tribes:

The Nevvahs, who never tell the truth;

The Oddfellas, who only tell the truth on days that are odd-numbered dates of the month;

The Pikkas, who only tell the truth on Wednesdays and Fridays;

The Quaints, who always tell the truth except on Mondays and Thursdays, when they make statements that are alternately true and false.

Adolf, Basil, Clyde, Dean and Elmer are five inhabitants of the island who make statements as follows:

ADOLF: (i) Today is the 14th.

         (ii) Basil is an Oddfella.

BASIL: (i) Today is the 13th.

         (ii) Elmer is a Nevvah.

       (iii) I am an Oddfella.

CLYDE: (i) Today is Monday.

        (ii) Elmer is a Pikka.

      (iii) Basil is a Pikka.

DEAN: (i) Basil is not a Quaint.

       (ii) Yesterday was the 15th.

     (iii) Clyde is a Pikka.

     (iv) Today is Monday.

ELMER: (i) Tomorrow is not Wednesday.

      (ii) Adolf is not a Quaint.

    (iii) Adolf belongs to the same tribe as Dean.

Find the tribe to which each person belongs, and find also the day of the week and the date of the month.

Answer on pages 93–94.

◆ Country Crescent

CASEY: Like you and me, Cuthbert, those three people all live in Country Crescent, and the numbers of their houses add up to twice the number of your house. Multiplied together, their three numbers make 1260. The numbers of our five houses are, of course, all different.

CUTHBERT: That doesn't tell me what their numbers are.

CASEY: That's true. But it will if I give you the additional information that the number of my house is greater than that of any of yours.

Country Crescent has houses numbered, rather curiously, from 2 to 222.

Find the numbers of the houses of Casey, Cuthbert and the three people they are talking about.

Answer on pages 94–95.

Solutions

◆ Predictions

1. Since the only one to predict correctly was the man who married Dana, John did not predict correctly (that Arthur would marry Dana). Therefore, Arthur did not marry Dana and John did not marry Dana.

2. Therefore, Arthur did not predict correctly. Therefore, David married Eve.

3. Therefore, by elimination, Dana married George. Therefore, George and only George predicted correctly.

4. Therefore, John did not marry Christine. Therefore, by elimination, Arthur married Christine and John married Rose.

Complete result: George married Dana; John married Rose; Arthur married Christine; David married Eve; Gallant Dancer did not win the race.

◆ The Architect's Shirt

	Red	Blue	Black	Green	Arch.	Photog.	Law.	Plumb.
Smith			×	×			×	
Brown					×			
Jones			×					×
Rodriguez								
Arch.	×							
Photog.	✔	×	×	×				
Lawyer	×		×	×				
Plumber	×			×				

1. Fill in the information given:
   a. The architect is not Brown.
   b. Smith is not the lawyer. Neither Smith nor the lawyer wore black or green shirts.

c. Jones is not the plumber. Neither Jones nor the plumber wore a green shirt.

d. The photographer wore the red shirt. The photographer wore no other color shirt, and no one else wore the red shirt.

(This information has been inserted in the diagram.)

2. From the diagram, by elimination, the architect wore the green shirt, the plumber wore the black shirt, the lawyer wore the blue shirt.

3. The architect wore green, but neither Smith nor Jones wore green. Therefore, the architect was not Smith or Jones. Therefore, the architect was Rodriguez, and since the architect wore green, Rodriguez wore green.

Complete results: Smith was the photographer and wore red; Brown was the plumber and wore black; Jones was the lawyer and wore blue; Rodriguez was the architect and wore green.

◆ Who Killed Popoff?

If A2 is true, then B2 is true, and then A2 is false. Therefore, A2 is false.

If B2 is true, then A and B have not made the same number of true statements. Therefore, B2 is false. Therefore, not everything that A says is false. Therefore, A (and B) must have made at least one true statement.

If A3 is true, then C2 is true, then B1 is false and B3 is false, and A and B have not made the same number of true statements. Therefore, A3 is false. Therefore, A1 is true (A must have made at least one true statement). Therefore, C1 is false.

If B3 is true, then C2 is false, then B1 is true; but this is impossible as A and B must make the same number of true statements. Therefore, B3 is false; therefore, B1 is true and C2 is false. Therefore, C3 is true (since B3 is false).

But, since A1 and B1 are both true, A and B must be innocent. Therefore, *Charles killed Popoff* (and some light is thrown on his third, carefully ambiguous remark).

◆ River Road

Their remarks, using obvious abbreviations, were:

A: The number of B's house is even.

B: 1. The number of my house is greater than the number of D's.

2. My age is a perfect cube.

C: 1. The number of my house is greater than the number of A's.

2. D's age is a multiple of A's age.

D. 1. B's age is either 27 or an even number other than 64.

2. C does not live at number 19.

Since difference between age and house number is always 7, therefore, if house number is even, age is odd; and if house number is odd, age is even. Therefore, false remarks are made by those with odd ages and even houses; true remarks are made by those with even ages and odd houses.

If B2 is true, B's age is 27 or 64, but if age is 27, the remark is false. Therefore, either B's age is 64 (and the remark true), or B's age is an odd number other than 27 (and the remark is false). D1 contradicts this; therefore, D1 is false.

Therefore, D's age is odd and his house number even, and D2 is false.

Therefore, C lives at number 19 and C's age is 26 (it cannot be 19 − 7).

Therefore, C's remarks are true, and from C1, A's house is number 16.

Therefore, A's age is 23 (it cannot be 16 − 7).

Therefore, A's remark is false.

Therefore, B's house number is odd and B's age is even.

But we know that B's age is either 64 or an odd number other than 27.

Therefore, B's age is 64.

Therefore, B's remarks are true.

Since C2 is true, D's age is a multiple of 23.

But D's age is odd (see above). Therefore, D's age is 69.

B1 is true. Therefore, the number of B's house is greater than the number of D's.

But B's age is greater than D's. B's house must be 64 + 7 (71), and D's house must be 69 − 7 (62).

Complete results: Andy is 23 and lives at number 16.

Brad is 64 and lives at number 71.

Cole is 26 and lives at number 19.

Doug is 69 and lives at number 62.

◆ Schoolmates

1. The biologist is not Catherine and not from Canada. Therefore, she belongs to C house (she must have one C).

2. If Catherine were the doctor, then she would have to come from Brazil (we know she's not from Australia,) and if she were the doctor, she couldn't be from Denmark. Therefore, the doctor is from Brazil—but we are told the doctor is not from Brazil. Therefore, Catherine is not the doctor. Therefore, she must be the author.

3. The biologist is not from Australia and not from Canada. Therefore, she must be from Denmark.

4. Since the biologist is from C house and from Denmark, she must be Alice.

5. Therefore, the doctor is not Alice, and the doctor is not Catherine. Therefore, the doctor is Brett and Deirdre must be the cartoonist.

6. Since the doctor is Brett, therefore, she is not from B house. Therefore, the doctor is from A house (the only alternative left). Therefore, the doctor is from Canada.

7. Since the cartoonist is Deirdre, she cannot be from D house. Therefore, she is from B house. Therefore, the cartoonist is from Australia.

8. Therefore, Catherine, the author, is from Brazil and was in D house.

The complete solution:

	Profession	House	Home
Alice	Biologist	C	Denmark
Brett	Doctor	A	Canada
Catherine	Author	D	Brazil
Deirdre	Cartoonist	B	Australia

◆ Vests and Vocations

	Bak.	Car.	Hunt.	Walk.	Brown	Crimson	Helio.	White
Ms. Baker	×				×			
Ms. Carpen.	✔	×	×	×		×		
Ms. Hunter	×		×	×			×	
Ms. Walker	×			×				×
Brown	×		×					
Crimson		×						
Helio.			×					
White								

1. The facts that Ms. Baker was not the baker, did not own a brown vest, etc., have been marked in.

2. Also marked in are the facts that Ms. Hunter was not the walker, that the hunter did not own a brown vest, and that Ms. Carpenter was the baker. (Note that from this last positive information we can deduce, as shown, that Ms. Carpenter was not the hunter or the walker, and that the baker was not Ms. Hunter or Ms. Walker.)

3. By elimination, Ms. Baker was the walker. Therefore, by elimination, Ms. Walker was the hunter. Therefore, by elimination, Ms. Hunter was the carpenter.

4. Since the walker was Ms. Baker, who did not own a brown vest, therefore the walker did not own a brown vest. Similarly, Ms. Baker did not own a white vest. By elimination, the carpenter owned a brown vest. Therefore, Ms. Hunter (who is the carpenter) owned a brown vest.

5. Therefore, by elimination, Ms. Carpenter owned a white vest. Therefore, the baker (Ms. Carpenter) owned a white vest.

6. Therefore, by elimination, the hunter owned a crimson vest, and, by elimination, the walker owned a heliotrope vest, and the rest follows easily.

Therefore, the complete solution is: Ms. Baker was the walker and owned a heliotrope vest; Ms. Carpenter was the baker and owned a white vest; Ms. Hunter was the carpenter and owned a brown vest; Ms. Walker was the hunter and owned a crimson vest.

◆ The Poison Spreads

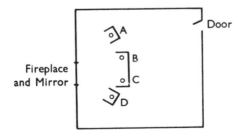

1. From (1) and (4), the headmaster is not Johnson or Bowen (mark in diagram as shown).

2. From (2), the general is not Robinson. And the general must be C or D (see first diagram) to see the door in the mirror. Robinson is not A.

	General	Headmaster	Admiral	Surgeon
Scott				
Bowen		×		
Johnson		×		
Robinson	×			

3. From (5), Scott is not the admiral or the headmaster. Therefore, by elimination, Robinson is the headmaster.

4. Also from (5), Scott is sitting at A, the headmaster at B.

5. Since the general is next to Robinson (2) and sitting at C or D, the general must be at C. Therefore, the general is not Scott. Therefore, by elimination, Scott is the surgeon. Therefore, by elimination, the admiral is sitting at D.

6. Since Scott is the admiral's brother-in-law, and neither Scott nor Bowen have any sisters, therefore the admiral is not Bowen. Therefore, Bowen is the general and Johnson is the admiral.

Complete results: A is Dr. Scott; B is Headmaster Robinson; C is General Bowen; D is Admiral Johnson. The hand putting something into Johnson's glass must belong to General Bowen.

◆ Hitler and Goering

```
    H I T L E R        _ _ _ _ _ _
    G O E R I N G      _ _ _ _ _ _ _
  H T T L L H H H      _ _ _ _ _ _ _ _
```

It is important to have a pattern, as above, and to fill in the blanks as you find them.

1. When two numbers are added together, the most that can be carried is 1. Therefore, H must equal 1.

2. Since the most that can be carried is 1, G must be 9 and T must be 0.

3. Therefore, H + 0 (and perhaps 1 carried from I + E) = 10. 0 cannot be 9 (G is 9). Therefore, 0 must be 8 and there *is* 1 carried from I + E.

4. Now look at the other end. G = 9, H = 1, R = 2.

5. Since R = 2 (below T), we have 0 + 2, apparently producing L. But L cannot be 2, therefore, there must have been a 1 to carry (from L + I) and L = 3.

6. Below L + I (3 + ?) we have H (1). I cannot be 8 (0 = 8). Therefore, I = 7 and there was 1 to carry from E + N.

7. Look now at the E of G O E R I N G, and it is easy to see that it must be 6. Therefore, N = 4.

Complete solution is:

```
    1 7 0 3 6 2
    9 8 6 2 7 4 9
  _____
  1 0 0 3 3 1 1 1
```

◆ Long Division #1

1. 5 and 8 are filled in, as shown, where they will be brought down, as is the figure to the right of 8 in *d* (it must be 0).

2. Since *b* subtracted from *a* produces a number less than ten, therefore, *a* must be 1005 and *b* greater than 995.

3. Since a multiple of b or a multiple of a factor of b ends in 0 (see d) and since b cannot end in 5 or 0, therefore b must end with an even digit. Therefore, b is 996 or 998.

4. Therefore, c is either 9– – – or 7– – –. Therefore, the divisor is greater than 700 (otherwise, the last figure in c would not have been brought down). Therefore, the divisor is 996 or 998.

5. d is five times the divisor. Therefore, the divisor is 996 (5 × 996 = 4980; 5 × 998 = 4990). Therefore, d is 4980.

6. c must start with 9. Therefore, the four-figure number below c is 8964 (996 × 9).

The rest follows easily!

```
                  1 0 0 1 0 0 9 0.5
      9 9 6 | 9 9 7 0 0 5 0 1 3 8
            9 9 6
            1 0 0 5
              9 9 6
                9 0 1 3
                8 9 6 4
                    4 9 8 0
                    4 9 8 0
```

◆ Long Division #2

1. Consider the first division. Three figures subtracted from four figures leaves one figure. Therefore, the divisor must go exactly into something between 991 and 999.

2. But from the second division, we see that the remainder from the first division is 1. Therefore, the divisor goes exactly into 999.

3. From the third division, the divisor multiplied by 9 or less produces four figures. Therefore, the divisor is not 111. Therefore, the divisor is 333 or 999.

4. Consider the third act of division (dividing into 1– –8. The biggest this could be is 1998. Therefore, the divisor is not 999 (2 × 999 = 1998 and there would be no remainder). Therefore, the divisor is 333. And the first two divisions must obviously be 999.

5. Consider the last act of division. 333 × x produces a number that, subtracted from y000, leaves something between 1 and 9. Possible values of x are 6 (333 × 6 = 1998) and 9 (333 × 9 = 2997). If x were 6, the last subtraction would be of 1998 from 2000. To produce 2, the previous multiple of 333 must end in 6 (bearing in mind that we are subtracting from 1– –8). The only possible multiple would be 333 × 2, but this only has three figures where there should be four. Therefore, x cannot be 6. Therefore, x must be 9. And the last subtraction must be of 2997 from 3000.

6. The figure to subtract from 8 to give 3 must be 5, and the penultimate division is 333 × 5 (1665). Add up from the bottom and the complete solution is:

```
                    3 0 0 3 0 5.0 0 9
    3 3 3 [ 1 0 0 0 0 1 5 6 8
            9 9 9
              1 0 1 5
              9 9 9
                1 6 6 8
                1 6 6 5
                    3 0 0 0
                    2 9 9 7
                        3
```

♦ Long Division #3

```
            - - -.- -
    - - | 1 0 - - -
        9 -
        ─────
        a a

         - -
        ─────
        - 3 -      (v)
        - - -      (iv)
        ─────
            - 0    (i)
            - b    (ii)
            ─────
            - 0  (iii)

            - -
            ───
```

1. When there are no more figures to bring down, a decimal point is put in the answer and 0's are brought down, as shown.

2. Since, in the first subtraction, the result of subtracting two figures from three figures is to produce one figure, therefore 9– must be subtracted from 10– (as shown).

3. Divisor cannot end in 0, for if it did, b would be 0, which would be absurd (the next line would be 00).

4. Divisor cannot be 9–, for the greatest value of (i) is 90, and (ii) would then be greater than (i).

5. If b were 5, the last division would be into 50, and the divisor could only be 25. But 25 does not go exactly into 9–, therefore the divisor does not end in 5. Therefore, since the bottom line ends in 0, the divisor must be multiplied by five; and since the bottom line has only two figures, therefore the divisor must be less than 20. The only possibilities are 12, 14, 16 and 18 (to go exactly into –0).

6. If the divisor were 14, (iii) would be 70 and b 3, which is impossible. If the divisor were 18, (iii) would be 90 and b 1, which is impossible.

7. If the divisor were 12, the greatest possible value of (iv)

would be 108 (12 × 9), and this subtracted from –3– must leave more than a single figure. Therefore, the divisor is 16.

8. Therefore, (iii) is 80; therefore b is 2 and (ii) is 32. Therefore, (i) is 40, (iv) must be 128 and (v) 132.

9. The multiple of 16, which subtracted from $a\,a$ leaves 13, can only be 64, and $a = 7$.

The rest follows easily and the final solution is:

```
                6 4 8.2 5
    1 6  1 0 3 7 2
             9 6
             7 7
             6 4
             1 3 2
             1 2 8
                 4 0
                 3 2
                   8 0
                   8 0
```

◆ Long Division #4

```
             - - -.- -
    - -  - - 5 - -
             - -          (i)
             - - -        (ii)
             - - -        (v)
             - b 0
               - - -      (iv)
             a - 0        (iii)
               - - -
```

1. When there are no more figures to bring down, a decimal point is put in the answer, and 0s are brought down, as shown.

2. Since (i) subtracted from ––5 leaves only a single digit, ––5 must be 105 and (i) must be between 96 and 99 inclusive.

3. (ii) must start with at least 6 (99 from 105). Therefore, the divisor must be greater than 6– (since two figures are brought down). Therefore, the divisor must be 96, 97, 98 or 99.

4. From (iii) the divisor goes exactly into ––0. Therefore, it must be 96 or 98. If 98, (iii) would be 490, and (iv) would end in 1. But no multiple of 98 ends in 1. Therefore, the divisor is 96. And since $5 \times 96 = 480$, therefore $a = 4$. Therefore, $b = 2$.

5. Since (iii) is 480, therefore (iv) is –72, and the only possible multiple of the divisor is 7×96 (672).

6. (ii) starts with 9 ($105 - 96$). Therefore, (v) must be 9×96 (864).

The rest follows easily:

$$
\begin{array}{r}
1\ 0\ 9.7\ 5 \\
\hline
9\ 6\ |\ 1\ 0\ 5\ 3\ 6 \\
9\ 6 \\
\hline
9\ 3\ 6 \\
8\ 6\ 4 \\
\hline
7\ 2\ 0 \\
6\ 7\ 2 \\
\hline
4\ 8\ 0 \\
4\ 8\ 0 \\
\hline
\end{array}
$$

♦ Long Division #5

```
        _ _ _._ _ _
_ _ _ | _ _ _ _ _
      a _ _
      _ _ a _
      _ _ _ _
      _ _ a  p
      _ _ _  q
          r _ _
          _ _ _
```

1. p must be 0. Therefore, q is not 0. If it were, the last two rows would be 000, which would be absurd. Therefore, the divisor does not end in 0.

2. Since the divisor goes exactly, less than 10 times, into -0 0, and does not end in 0, it must end in 5. Therefore, q is 5, and r is 5.

3. It is easy to see that the divisor must be 125.

4. $8 \times 125 = 1000$. Therefore, the four-figure multiple ending in 5 can only be $9 \times 125 = 1125$.

5. Therefore, $--ap$ is 1125 (i.e., 1130). Therefore, $a = 3$.

6. Therefore, the first multiple of 125 must be 375.

The rest follows easily. The complete solution is:

```
              3 0 9.9 0 4
    1 2 5 | 3 8 7 3 8
          3 7 5
          1 2 3 8
          1 1 2 5
            1 1 3 0
            1 1 2 5
                5 0 0
                5 0 0
```

◆ Long Division #6

```
          _._ _
  _ _ |‾‾‾‾‾‾‾
  _ _ |_ _ _
       _ _        (p)
      ‾‾‾‾‾
       d 0
       c a
       ‾‾‾‾
       b 0
       _ _
       ‾‾‾‾
```

1. Both figures brought down must be 0, as marked.

2. a cannot be 0. If it were, b would be 0, which would be absurd. Therefore, the divisor cannot end in 0.

3. If the divisor ends in 5, then $a = 5$ and $b = 5$. Therefore, the divisor would be 25. But if the divisor is 25, it would go at least four times at the first act of division, and (p) would contain three figures instead of two. Therefore, the divisor cannot end in 5.

4. For the divisor to go exactly into $b0$, it must end in 2, 4, 6 or 8, and it must go into $b0$ five times. Therefore, the divisor must be 12, 14, 16 or 18.

5. If the divisor were 14, $b = 7$ and $a = 3$, which is impossible.

6. If the divisor were 18, $b = 9$ and $a = 1$, which is impossible.

7. If the divisor were 12, $b = 6$ and $a = 4$. ca would then have to be 24 or 84. If ca were 24, d would be 3, and since p is 96, the dividend would be 99. But the dividend contains three figures. Therefore, this is impossible.

If ca were 84, d would be 9, and a solution is possible:

$$
\begin{array}{r}
8.7\ 5 \\
1\ 2\ \overline{\smash{\big)}\ 1\ 0\ 5} \\
\underline{9\ 6} \\
9\ 0 \\
\underline{8\ 4} \\
6\ 0 \\
\underline{6\ 0}
\end{array}
$$

8. If the divisor were 16, $b = 8$, $a = 2$, $c = 3$ and $d = 4$. And the complete solution is:

$$
\begin{array}{r}
6.2\ 5 \\
1\ 6\ \overline{\smash{\big)}\ 1\ 0\ 0} \\
\underline{9\ 6} \\
4\ 0 \\
\underline{3\ 2} \\
8\ 0 \\
\underline{8\ 0}
\end{array}
$$

◆ Add and Subtract #1

(i) d q c (ii) d q c
         p b p b
     ——————   ——————
         s a e r a

Obviously, (i) is subtraction and (ii) is addition. Since $b - c = 1$, therefore, from (i), $a = 9$.

Since b added to c and b subtracted from c produce the same digit, and there are no 0's, therefore $b = 5$ and $c = 4$.

From (i), $d = 1$ (if 2 or more, the result of the subtraction would be more than 100). In (ii), since e is not 1 (figures all different), therefore $e = 2$.

From (i), p is greater than q; from (ii), $p + q$ is greater than 12. (There is not one to carry from $b + c$, and r cannot be 0, 1 or 2.)

If p is 6 or less, we cannot have p greater than q, and $p + q$ is at least 12. Therefore, p can only be 7 or 8 ($a = 9$).

If $p = 8$, $q = 6$ or 7 (not 5 because $b = 5$).

If $p = 8$ and $q = 6$, then $r = 4$; but $c = 4$, and we are told that no two letters in (ii) are the same.

If $p = 8$ and $q = 7$, then $r = 5$; but $b = 5$. Therefore, p is not 8. Therefore, p is 7. Therefore, $q = 6$ (p is greater than q and $[p + q]$ is greater than 12). Therefore, $r = 3$ and $s = 8$.

Complete solution is:

(i)
```
    1 6 4
  −   7 5
  ─────────
      8 9
```

(ii)
```
    1 6 4
  +   7 5
  ─────────
    2 3 9
```

◆ Add and Subtract #2

(i)
```
  c   d   e
      a   b
─────────────
g   h   j   k
```

(ii)
```
  c   d   e
      a   b
─────────────
  b   l   a
```

Obviously, (i) is the addition and (ii) is the subtraction.

The most that can be carried when two numbers are added is 1. Therefore, $q = 1$. Hence, $c = 9$ and $h = 0$. From (ii), since c and b are different and $c = 9$, therefore $b = 8$.

From (ii), a is greater than d, and from (i), $a + d$ is greater than 10 (there might be 1 to carry from $e + b$, but j cannot be

1). Bearing in mind that different letters stand for different digits, a little trial and error will show that $a = 7$ and $d = 4$. Since $b = 8$, therefore $b + e$ is greater than 10. Therefore, $j = 2$ and $l = 6$. It follows easily that $e = 5$ and $k = 3$.

Complete solution is therefore:

(i)
```
  9 4 5
    7 8
1 0 2 3
```
(ii)
```
9 4 5
  7 8
8 6 7
```

◆ Long Division #7

1. Divisor goes exactly into $x00$ (see end) less than 10 times.

2. If the divisor ended in 0, every multiple of it would end in 0, and d would have to be equal to a. We are told this is not so. Therefore, the divisor does not end in 0. Therefore, the divisor ends in 5 (otherwise, a single-figure multiple of it could not be $x00$).

3. It is easy to see that the divisor must be 25 or 75 to go exactly into $x00$. If 25, the highest possible value of the first division would be 75. Subtracted from the three-figure number, this would leave 25 or more—which is absurd. Therefore, the divisor is not 25. Therefore, the divisor is 75.

4. Therefore, each two-figure multiple of the divisor is 75.

5. From a in the quotient, we see that $a = 1$.

6. Fill in other a's, and from the fact that each two-figure multiple is 75, we see that $d = 6$.

7. Three figures below a— must be 150 (since $a = 1$).

8. Since $b - c = 2$, and since $bc-$ must be a multiple of 75 plus 7, it is easy to see that $bc-$ must be 532 (525 + 7).

The solution is:

```
                    1 7 1 0 2 1 1.0 8
        7 5 ⟌ 1 2 8 2 6 5 8 3 1
              7 5
              5 3 2
              5 2 5
                  7 6
                  7 5
                    1 5 8
                    1 5 0
                        8 3
                        7 5
                          8 1
                          7 5
                            6 0 0
                            6 0 0
```

◆ Addition and Division

1. Since the most that can be carried from adding two rows is 1, the four-figure number must start with 99, and the result of the addition must start with 100.

```
        _ _ _ _ _
  _ _ ⟌ 9 9 _ _ 0 0
        _ _                (i)
        a  c  _
        _ _
            _ 0 0
            _ _ _

              _
```

66

2. *a* can only be 1 (a two-figure number subtracted from *a c* – leaves less than 10). Therefore, (i) must be 98. Therefore, the divisor must be 98 or a factor of 98.

3. *c* must be 0 (98 subtracted from 1 *c* – leaves less than 10).

4. The addition now looks like this:

$$
\begin{array}{r}
9\ 9\ 0\ - \\
d\ - \\
\hline
1\ 0\ 0\ -\ -
\end{array}
$$

In order that there shall be 1 to carry from the column in which *d* is, *d* must be 9. Therefore, the divisor must be 98.

5. The division now looks like this:

$$
\begin{array}{r}
\overline{-\ -\ -\ -\ -} \\
98\ |\ 9\ 9\ 0\ e\ 0\ 0 \\
9\ 8 \\
\hline
1\ 0\ - \\
9\ 8 \\
\hline
-\ 0\ 0 \\
-\ -\ - \\
\hline
\end{array}
$$

$$\overline{}\ -$$

and the addition like this:

$$
\begin{array}{r}
9\ 9\ 0\ e \\
9\ 8 \\
\hline
1\ 0\ 0\ 0\ -
\end{array}
$$

6. Consider the addition: *e* must be at least 2 to produce 1 to carry.

7. Consider the division: suppose *e* is 3. Then the last act of division would be 98 into 500, which would go 5 times and leave a remainder of 10. But there is only one digit in the remainder. Therefore, *e* cannot be 3 (even less can it be greater than 3). Therefore, *e* = 2, and the rest follows easily.

```
        1 0 1 0 4
9 8 | 9 9 0 2 0 0
      9 8
      1 0 2
        9 8
          4 0 0
          3 9 2
              8
```

$$9\,9\,0\,2$$
$$\underline{9\,8}$$
$$1\,0\,0\,0\,0$$

◆ The Christmas Compensation Club

Use initial capitals for their ages. Facts given are:

$$(K - I) = 3(K - H) \tag{i}$$

$$\frac{J}{H} = \frac{9}{10} \tag{ii}$$

$$\frac{J}{I} = \frac{6}{5} \tag{iii}$$

$$G - K = J - I \tag{iv}$$

Since all letters must stand for whole numbers, therefore from (ii) and (iii) J must be a multiple of 9 and of 6. Therefore, a multiple of 18 (denote by m(18). Therefore, H is m(20), and I is m(15).

From (i), $2K = 3H - I$. (v)

Therefore, H and I must be both odd or both even (for 2K must be even).

But H is m(20) and cannot be odd. Therefore, H and I are both even. Therefore, H is m(40) and I is m(30); therefore, J is m(36).

If H is 80, I 60 and J 72, then from (v) 2K = 180; therefore, K = 90.

But we are told that none of them has reached the age of 90. Therefore, *H = 40, I = 30, J = 36, K = 45*, and from *(iv) G = 51.*

◆ The Years Roll By

1. Consider B's remark. If true, it is false. (He cannot both be 81 and be making a true statement.)

2. Therefore, C's remark is false. Therefore, C is under 50 or 64 or 81.

3. D (ii) is telling the truth. Therefore, D is over 50 or 27 and D's (i) is true. Therefore, A is 57. Therefore, A is truthful, and therefore, E is 27. Therefore, E is truthful, and therefore, B is older than A.

But B is under 50 or 64 [see (i)] and A is 57. Therefore, B is 64.

4. E's (ii) is true. D is 27 or over 50; C is 64, 81 or under 50, and D is 30 years younger than C. Therefore, D is 51 and C is 81.

Ages are: Annie, 57; Bill, 64; Chris, 81; Dirk, 51; and Erika is 27.

◆ Sinister Street

1. For A to claim that she knows where X lives, she must certainly think his number is a perfect square. If she thinks it's a perfect square less than 50, there are too many alternatives (1, 4, 9, 16, etc.). But there are only *two* squares between

50 and 99—namely 64 and 81. Clearly, the only way in which A can claim to know X's number is if she lives in one of these houses and thinks X lives in the other. Therefore, A's number is 64 or 81. X's number is *not* a square greater than 50.

2. Since X answers B's second question truthfully, X must say his number is greater than 25. And for B to claim to know X's number, she must certainly think it's a perfect cube. Therefore, B must think it's a perfect cube. Therefore, B must think X's number is 27 or 64, and she can only claim to know which if she lives in one of those houses herself. Therefore, B's number is 27 or 64. X's number is not a cube.

3. Since X's number is greater than 50 and less than that of A or B, therefore A's number must be 81, B's must be 64 and X's must be between 51 and 63. The sum of their three numbers is a perfect square multiplied by two. It is easy to see that this must be 200. Therefore, X's number is 55.

Therefore, *Agnes lives at #81; Beatrice lives at #64; Xerxes lives at #55.*

◆ Youthful Ambitions

Denote the names of the men by A1, B1, etc., the wives by A2, B2, etc., the birthplaces by A3, B3, etc., ambitions by A4, B4, etc. A diagram will help.

1. Since, for each man there are four different initial letters, A1, A2, B1, B3, etc., are crossed out, as shown.

2. All E2's remarks are true (see conditions). Therefore, from E2's second comment, C3 is E4 (marked in diagram). Therefore, C3 is not A4, B4, etc., and E4 not A3, B3, etc.

Since C3 is E4 and the letters are all different, C3 is not E1 or E2, and since E4 is C3 and the letters are all different, E4 is not C1 or C2.

	A2	B2	C2	D2	E2	A3	B3	C3	D3	E3	A4	B4	C4	D4	E4
A1	×					×					×				
B1		×				×	×					×			
C1			×			×		×					×		×
D1	×			×		✔	×	×	×	×	×			×	
E1					×	×		×		×					×
A4	×					×		×							
B4		×					×	×							
C4			×					×							
D4				×		×		×	×						
E4			×		×	×	×	✔	×	×					
A3	×			×											
B3		×													
C3			×		×										
D3				×											
E3					×										

3. From E2's first comment, which is true, A3 is not B1 and A3 is not E1 (since remark is made by E2).

4. Consider C2's remark. The subject, A3, is not C1.

5. Therefore, by elimination, A3 is D1. Therefore, A3 is not D2 or D4; D1 is not A2 or A4.

(Information until now has been marked in the diagram.)

6. Since A3 is D1, therefore C's remark is false. Therefore, A3 is not B2.

7. A2's remark is false (see conditions). Therefore, D4 is not B2, and D4 is not A1.

8. From B2's first comment, E4 is not B1. From D's remark, B4 is not D1.

9. A3 is D1. Since A3 is not E4, therefore D1 is not E4. Since A3 is not B2, therefore D1 is not B2.

10. By elimination, D1 is C4. But D1 is A3; therefore, A3 is C4. Since D1 is C4, therefore D1 is not C2. Therefore, by elimination, D1 is E2. But D1 is A3 and C4; therefore, E2 is A3 and C4.

11. By elimination, E4 is A1. Therefore, B2's first comment is true. Therefore, E4 is not D2. Since E4 is A1, therefore E4 is not A2; therefore, E4 is B2. But E4 is A1; therefore, B2 is A1, and E4 is C3. Therefore, A1 is C3.

12. A1 is B2; A1 is C3; therefore, B2 is C3.

13. Consider B2's second comment: (B3 is D4). We know B3 is not A1 (C3 is). Therefore, the comment is false. Therefore, B3 is not D4. Therefore, by elimination, B3 is A4. Therefore, A4 is not B1. Therefore, by elimination, D4 is B1.

14. Therefore, B1 is not D2 or D3. Therefore, by elimination, B1 is E3. But B1 is D4, therefore E3 is D4.

15. Therefore, by elimination, B4 is D3. Therefore, B4 is not D2. Therefore, by elimination, A4 is D2. But A4 is B3, therefore D2 is B3.

16. Consider D2's remark. We know that this is true (B4 is *not* D2). Therefore, B4 must be A1, B1 or C1. But, of these, only C1 is possible. Therefore, B4 is C1. Therefore, B4 is not C2. Therefore, by elimination, B4 is A2 and therefore C1 is A2.

17. Therefore, by elimination, D2 is E1, and C2 is B1. D4 is C2. B1 is E3; therefore, C2 is E3. Therefore, by elimination, A2 is D3, and therefore C1 is D3. Therefore, E1 is B3 and E1 is A4.

Complete solution:

Adam	*Barbara*	*Chicago*	*Egyptologist*
Biff	*Chita*	*Edmonton*	*Dog breeder*
Chad	*Alison*	*Delhi*	*Banker*
Dan	*Erin*	*Amherst*	*Chemist*
Earl	*Danielle*	*Bristol*	*Astronaut*

◆ Uncle Knows Best

We are clearly only interested in those answers to (i), (ii) and (iii), which reduce the possible alternatives to a small number. These are:

	(i)	(ii)	(iii)	Numbers
(a)	Yes	Yes	Yes	36
(b)	Yes	Yes	No	16, 64, 100
(c)	Yes	No	Yes	72
(d)	No	Yes	Yes	9, 81
(e)	No	Yes	No	25, 49

1. Consider B's remark to X. B must think that X's number is one of two, one of which is greater than 83, but not the other. The only category in which this is possible is b. But here there are two alternatives less than 81. The only possibility is for B to be one of them. Therefore, B's number must be 16 or 64; and he must think that X is the other of these—or 100. C hears, and knowing B's number as well as his own, is able to write down X's number correctly. This is only possible if C himself lives in 16, 64 or 100.

2. Consider B's remark to Y. B must think that Y's number is one of two, one of which is greater than 50, but not the other. Remembering that no two sets of answers are exactly the same, this can only be in category d. Therefore, Y's number is 9 or 81. C knows that Y's number is greater than his (and we know that C's number is 16, 64 or 100). Therefore, Y's number must be 81, and C's number must be 16 or 64.

3. Consider B's remark to Z. His number must be one of two, one of which is greater than 30, but not the other. Since no two sets of answers are exactly the same, this can only be in category e. Therefore, Z's number is 25 or 49. C notes that his number is less than Z's. Therefore, C's number must be 16 and not 64. B's number is 64, and X's number is 100. And

C guesses, correctly, that Z's number is greater than 30. Therefore, Z's number is 49.

Complete solution: Bernard = 64; Chester = 16; X = 100; Y = 81; Z = 49.

◆ Out the Window and Over the Wall

Select the times at which the sentries are out of range—more than 100 yards away.

P1	Q1	R1	S1
9:00–9:04	9:00½–9:09½	9:00–9:02	9:01–9:07
9:07–9:15	9:10½–9:19½	9:04–9:08	9:09–9:15
9:18–9:26	9:20½–9:29½	9:10–9:14	9:17–9:23
9:29–9:37	9:30½–9:39½	9:16–9:20	9:25–9:31
9:40–9:48	9:40½–9:49½	9:22–9:26	9:33–9:39
		9:28–9:32	9:41–9:47
		9:34–9:38	
		9:40–9:44	

We have to find the first three periods of two minutes or more that are common to all of them.

They are: (1) 9:10½ until 9:14
(2) 9:34 until 9:37
(3) 9:41 until 9:44.

◆ The Willahs and the Wallahs

1. C to E: "I'm a Willah." If true, this is spoken by a Willah to a Willah. If false, it is spoken by a Wallah to a Willah. In both cases, E must be a Willah.

2. E to B: "My number is 35." Since E is a Willah, whose numbers are prime, this is false. Therefore, B is a Wallah.

3. E to A: "I'm a Wallah." This is false. Therefore, A is a Wallah.

4. A to B: "D is a Wallah." Since A and B are both Wallahs, this is true. And since there are three of each tribe, A, B and D are Wallahs; C, E and F are Willahs.

5. F to C, comment must be true. Therefore, F − C = 10.

6. B to D, comment must be true. Therefore, F's number is halfway between D's and C's. Therefore, D − F = 10.

7. C to F, comment must be true. Therefore D's number is halfway between A's and F's. Therefore, A − D = 10, and D − F = 10, and F − C = 10.

And the numbers of A, D, F and C must all be less than 50 (see D to A). A's and D's are *not* primes; F's and C's are primes. Therefore, the numbers of A, D, F and C are such that each is greater by 10 than the following one, and all are less than 50. A little investigation shows that the only possibilities are 49, 39, 29 and 19, in that order.

Complete solution is therefore:

   A: Wallah (not prime) = 49.
   B: Wallah (not prime) = ?
   C: Willah (prime) = 19.
   D: Wallah (not prime) = 39.
   E: Willah (prime) = ?
   F: Willah (prime) = 29.

The exact numbers of B and E cannot be discovered.

◆ Tom, Dick and Harry

The first answer from Dick cannot be true (if true, it is false).

Therefore, Dick is not a Wotta-Woppa, but Dick is not a Pukka, because he has made a false statement.

Therefore, Dick is a Shilli-Shalla, and Dick's second answer is true. Therefore, Tom is a Pukka.

Therefore, Harry's second answer is false. Therefore, Harry's first answer is false. Therefore, Harry is a Wotta-Woppa.

Therefore, Tom is a Pukka, Dick is a Shilli-Shalla, Harry is a Wotta-Woppa.

◆ Monogamy Comes to the Island

(i) If F2 is true, F must be SS (not WW, because she has made a true remark; not P, because two Ps cannot be married to each other).

Therefore, F1 is false. Therefore, G is a P. Therefore, G1 is true. Therefore, H2 is false. Therefore, H1 is true (from G1, H is not a WW); therefore, F is a P. But this is contrary to hypothesis (F must be SS). Therefore, hypothesis is false. Therefore, F2 is false. Therefore, F is SS or WW. Therefore, H1 is false. Therefore, H is SS or WW.

(ii) If G1 is true, then H is not a WW. Therefore, H2 is true. Therefore, G1 is false, which is contrary to hypothesis. Therefore, the hypothesis is false. Therefore, G1 is false. Therefore, F1 is true (since G makes at least one false remark). Therefore, F is an SS (F1 is true, F2 is false). Therefore, F3 is true, and therefore F's husband is Tired.

(iii) Since F is an SS, therefore T is not an SS. And, since F2 is false, therefore T is not P; therefore, T is WW.

(iv) G3 is false (because G1 is). Therefore, E is not WW, and H3 is false (because H1 is). Therefore, E is not SS; therefore, E is P and all E's statements are true.

(v) S is married to G (E1); S is WW (E2); therefore, G is not WW. Therefore, G is SS (we know that G1 and G3 are false).

(vi) Therefore, G2 is true and H2 is false. Therefore, H is WW.

(vii) Since G2 is true, therefore U is P. Therefore, U is not married to E. Therefore, U is married to H, and V is married to E.

(viii) Since E3 is true, therefore V must be an SS.

Complete result:

Eager (P) is married to Vulgar (SS).

Friendly (SS) is married to Tired (WW).

Glorious (SS) is married to Sordid (WW).

Happy (WW) is married to Ugly (P).

◆ Uncles and Cousins

From B(1) and B(2), we can draw the following diagram:

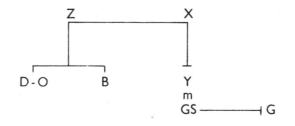

Z is B's father, X is Z's brother, Y is X's son and G's brother-in-law. And GS stands for Greg's sister. (This diagram accounts for six of the seven people.)

Consider now the following remarks:

    G: B is the D-S's cousin.

  E(2): My father is the W.

    F: My nephew is the D-S.

The only combination that will fit these facts and the diagram is for Y to be D-S, for X to be E, for E's father (not represented in the diagram, so that he must be the seventh man) to be W, and for Z to be F.

Our diagram now looks like this:

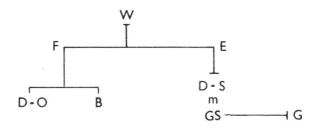

Consider E(1): There are two sets of brothers, E and F, and D-O and B. D and D-P cannot be E and F. Therefore, D must be D-O, and B must be D-P.

From D(1): F must be B-W (for F is D's father).

From A(1): A must be D-S and G must be CC (there are no other brothers-in-law).

By elimination, E must be S-U, and C must be W.

Complete solution is therefore:

Alfred —Door-Shutter

Barry —Doorknob-Polisher

Carl —Worker

Donald —Door-Opener

Edward —Sweeper-Upper

Ferdie —Bottle-Washer

Greg —Chief Cook

Carl is the father of Edward and Ferdie. Ferdie has two sons, Barry and Donald. Edward's son, Alfred, is married to Greg's sister.

◆ Intellectual Awareness

Consider A's remark. He is higher than B, and his place and B's place must be such that it is possible for C to be three places higher than D in one way only; and then of course E's is the remaining place.

The possible positions for A and B so that C is three places higher than D are as follows:

(i)   1 and 3 (C2, D5)   (iv) 2 and 5 (C1, D4)

(ii) 1 and 4 (C2, D5)   (v)   3 and 4 (C2, D5)

(iii) 2 and 3 (C1, D4)   (vi) 3 and 5 (C1, D4)

It is therefore certain that A's and B's places must be one of (i) to (vi). But remember that we have no information that C is in fact three places higher than D.

E is also able to deduce that A's and B's places are one of (i) to (vi). In addition, he knows his own place and assumes (correctly) that B is higher than D. As a result, he is able to deduce everyone's place.

Note that 3 appears in four of the possibilities for A and B in the chart above. No other number comes in more than two of them. If E were first, (i) and (ii) would be ruled out, also (iv) and (vi) (B is higher than D, therefore B is not fifth), but the following are all possible: E A B C D; E A B D C; E C A B D.

Similarly, it can easily be seen that if E were second, fourth or fifth, there are at least two possibilities.

But if E were third, the only possibilities are (ii) and (iv). (iv) is ruled out because B must be higher than D. E would therefore know with certainty that A and B must be first and fourth. And since B must be higher than D, he would know that D is fifth, and, therefore, C is second. These are the only conditions that make it possible for E to arrive, as he does, at the correct conclusion.

The order is, therefore, A C E B D.

◆ An Intelligence Test

Diagrams will help.

1.	A B C D E
P	x x
Q	x x
R	
S	
T	

2.	Places	
1.	BDE, *not* AC	PQRST
2.	ABCDE	PQRST
3.	ABCDE	PQRST
4.	ABCDE	PRST, *not* Q
5.	BDE, *not* AC	PRST, *not* Q

In diagram 1 we can insert information about who is or who is not P, Q, R, S and T. In diagram 2 we can insert information, often negative, that, for example, C is not first, Q is not fourth, etc.

1. From P's remark, P is not B or C. We cannot tell whether P is above B and C or below them.

2. From Q's first remark, Q is not A or C. If Q's remark is false, either A or C was fifth. Therefore, Q is above one of them and, therefore, above both. Therefore, Q's remark is

true, which is contrary to the hypothesis. Therefore, Q's first statement is true. Therefore, neither A nor C was fifth, and Q was above A and C. Therefore, Q is not third, fourth or fifth, and neither A nor C was first.

(Information so far has been inserted in the diagrams.)

3. From Q's second remark, Q is not E. (We cannot tell yet whether the statement is true.)

4. From R's remark, R is not C or E.

5. From S's remark, S is not D. If it were true, S would have to be above D, which is impossible. Therefore, D is not first. And S must be below D; therefore, D is not fifth and S is not first or second.

6. From T's first statement, T is not E. From T's second statement, T is not C or A. Therefore, by elimination, S is C, P is E, and R is A.

7. From diagram 2, we can now see that only B or E can be first, and only B or E can be fifth. Therefore, B and E between them are first and fifth, and they cannot be second, third or fourth.

8. Therefore, T's first remark is false. Therefore, T is below E. Therefore, E is not fifth. Therefore, E is first and B fifth.

9. We know that E is P. Therefore, P is first.

10. We know that B is fifth and is Q or T, but Q is not fifth. Therefore B is T, and T is fifth; and, by elimination, Q is D.

11. Q is not fourth; Q is D. Therefore, D is not fourth. S is not second, S is C. Therefore, C is not second.

12. Q is first or second (see ii). But P is first; therefore, Q is second. Therefore, D is second.

13. R's remark is false (E is first). Therefore, R is below C (as well as below E). Therefore, R is below S (who is C). Therefore, R is fourth and S is third. Therefore, A is fourth and C is third.

Solution: 1. Ernie (P); 2. Des (Q); 3. Candice (S); 4. Annette (R); 5. Bonnie (T).

◆ Birthdays

1. C to A (i): If true, this is false. Therefore, it must be false. Therefore, C is older than A, but not ten years.

2. C to A (ii): This is false (see above). Therefore, D is younger than B.

3. D to B: We know D is younger than B. Therefore, this is true. Therefore, D is nine years older than E.

4. From (2) and (3), we know E is younger than B. Therefore, E to B is true. Therefore, E is seven years older than A.

5. C to D: If C is six years younger than D, then this is true, and C is ten years older than A $(9 + 7 - 6)$. But we know this is not true [see (i)], therefore C is not six years younger. If C is six years older than D, this would have to be false. Therefore, the remark is false. Therefore, C is older than D, but not six years.

6. B to C (i): We know this is true. Therefore, B is younger than C.

7. We now know the order of ages, and some of the differences.

Thus, C, B, D ← 9 years → E ← 7 years → A.

8. B to C (ii) is true [see (6)]. Therefore, C is nine years older than D. Therefore, B is between 16 $(9 + 7 + 9)$ years older than A.

9. Since A to B is true, therefore:

$$\frac{\text{age of B}}{\text{age of A}} = \frac{170}{100} = \frac{17}{10}$$

Therefore, the age of B is a multiple of 17; the age of A is the same multiple of 10; therefore, the difference between their ages is a multiple of seven, and we know it is between 16 and 25. Therefore, it is 21. Therefore, B is 51 and A is 30. The rest follows easily.

Ages are: A, 30; B, 51; C, 55; D, 46; E, 37.

◆ Easter Parade

A diagram will help.

	WIVES							WIVES' HATS						
	A	B	C	D	E	F	G	A	B	C	D	E	F	G
True or false			×		×									
HUSBANDS A	×			×				×						
HUSBANDS B	×	×	×	✔	×	×	×		×		×			
HUSBANDS C			×	×						×				
HUSBANDS D				×							×			
HUSBANDS E				×	×							×		
HUSBANDS F				×		×							×	
HUSBANDS G			×	×			×							×
WIVES' HATS A	×													
WIVES' HATS B		×		×										
WIVES' HATS C			×											
WIVES' HATS D				×										
WIVES' HATS E					×									
WIVES' HATS F						×								
WIVES' HATS G							×							

   1. The information that name, wife's name and wife's hat begin with different letters has been filled in.

   2. Consider F's second remark. Either Fifi or her husband, *but not both*, is telling a lie. Therefore, Deb is Barry's wife. (Fill in this information in the diagram.)

   3. It also follows that Deb's hat is not B, and that Barry's wife's hat is not D.

   4. Therefore, C1 is false. Therefore, C is a liar. Therefore, her husband is not Greg (C2). Therefore, C3 is false. Therefore, Enid's husband is not a liar. Therefore, Enid is a liar. (All this information has been filled in.)

   5. Therefore, from E1, Alfred is a liar. Therefore, Alfred is not the husband of C or E (both liars).

   6. From E2, C's wife is wearing an E hat. Therefore, C's wife is not Enid. Therefore, the E hat is not being worn by Cleo (Cleo cannot be C's wife). Also C's wife is not Deb

(Bert's wife). Therefore, the E hat is not being worn by Deb.

7. E3 is false. Therefore, Cleo said *Yes*. But Cleo is a liar. Therefore, she is not Edward's wife.

8. B1 is true (see diagram; we know that Alfred is not married to Enid). Therefore, all B's remarks are true. Therefore, B2 is true. Therefore, C's husband did say *Yes*. We know that Cleo is a liar, therefore her husband tells the truth; therefore, *Ferdie is married to Gay*. Therefore, Gay cannot wear F hat, or the E hat (worn by Carl's wife), and as Ferdie's wife, does not wear G hat.

9. By elimination, Cleo is married to Donald; therefore, Donald's wife does not wear C hat, and Cleo does not wear D hat.

10. By elimination, Enid is married to Greg; therefore, Greg's wife does not wear E hat (already known), and Enid does not wear G hat.

11. *Greg and Donald are truthful* (married to Enid and Cleo).

12. Consider A1. We know G is *not* wearing F hat (see #8) and we know that Greg is truthful (#11); therefore, Amy is truthful.

13. Consider A2. We know Amy and Donald are both truthful; therefore, Edward's wife is wearing a B hat. Therefore, E's wife is not B.

14. Consider D1. Donald *is* married to Cleo. But Alfred is a liar. Therefore, Alfred could not have said it; therefore, Deb is a liar. Therefore, Barry, her husband, is truthful.

15. Therefore, D2 is false. Therefore, Edward is truthful. Therefore, Edward is not married to Amy (also truthful). Therefore, by elimination, *Edward is married to Fifi*, and by elimination, Carl is married to Amy, and by elimination, Alfred is married to Beth.

16. Fifi is a liar (married to truthful Edward). Therefore, F1 is false. Therefore, Donald's wife (Cleo) is not wearing A hat.

17. Fifi is wearing B hat (see #13 and #15). We know that Barry is truthful. Therefore, G1 is true. Therefore, *Gay is truthful*. Therefore, Ferdie lies.

18. Therefore, G2 is true. Therefore, Barry's wife (Deb) wears A hat.

19. D3 is false. Therefore, Ferdie's wife (Gay) does not wear C hat. Therefore, by elimination, Gay is wearing D hat.

20. Beth is married to Alfred (liar). Therefore, Beth is truthful. Therefore, B3 is true. But Ferdie is a liar. Therefore, Greg's wife (Enid) is not wearing F hat. Therefore, by elimination, Enid is wearing C hat.

21. Consider A3. Amy is truthful. Therefore, Cleo did say *Yes*. But Cleo lies. Therefore, Beth is not wearing F hat. (Note that Carl's wife, A, is wearing E hat.) Therefore, by elimination, Beth is wearing G hat, and Cleo F hat.

Complete solution:
Alfred (liar) married to Beth (truthful) who wears G hat.
Barry (truthful) married to Deb (liar) who wears A hat.
Carl (liar) married to Amy (truthful) who wears E hat.
Donald (truthful) married to Cleo (liar) who wears F hat.
Edward (truthful) married to Fifi (liar) who wears B hat.
Ferdie (liar) married to Gay (truthful) who wears D hat.
Greg (truthful) married to Enid (liar) who wears C hat.

◆ Trips Abroad

A diagram will help.

		WIVES					PLACES				
		A	B	C	D	E	a	b	c	d	e
HUSBANDS	A	×				×	×				
	B		×			×		×			
	C	×	×	×	×	✔			×		
	D				×	×				×	
	E					×					×
PLACES	a	×									
	b		×								
	c			×							
	d				×						
	e					×					

Since of each married couple, one member is truthful and the other lies, what anyone says that his or her partner says must be false. Therefore, from what Crystal says, Emily is married to Curt.

(This information has been inserted in the table.)

Therefore, Curt and Emily didn't go to Copenhagen or Ethiopia.

But Darcy said that Curt went to Ethiopia. Therefore, Darcy is a liar. Therefore, Brigitte did not go to Dublin and Anthony is a liar.

And since Anthony and Darcy are both liars, they are not married to each other. Brigitte said she was not married to Anthony. If Brigitte *were* married to Anthony, she would have to tell the truth (for Anthony is a liar). Therefore, she couldn't say she was *not* married to Anthony if she were. Therefore, she is not married to Anthony, and Brigitte tells the truth.

By elimination, *Anthony is married to Crystal*. Therefore, Anthony and Crystal didn't go to Auckland or Copenhagen. Since Anthony is a liar, therefore Crystal tells the truth.

Therefore, Derek went to Bologna. Therefore, Derek is not married to Brigitte and Darcy did not go to Bologna.

By elimination, *Ellery is married to Brigitte*, and therefore they didn't go to Bologna or Ethiopia.

By elimination, *Derek is married to Ava*, and since Derek went to Bologna, therefore Ava did too.

By elimination, *Ben is married to Darcy.* Therefore, they didn't go to Bologna or Dublin.

Since Brigitte tells the truth and is married to Ellery, therefore Ellery is a liar. Therefore, Curt did not go to Dublin.

Therefore, by elimination, *Curt went to Auckland.*

Therefore, *Emily (Curt's wife) went to Auckland.*

Therefore, by elimination, *Brigitte went to Copenhagen.*

Therefore, *Ellery went to Copenhagen.*

By elimination, *Ben and Darcy went to Ethiopia.*

By elimination, *Anthony and Crystal went to Dublin.*

Complete solution: Anthony and Crystal went to Dublin; Ben and Darcy went to Ethiopia; Curt and Emily went to Auckland; Derek and Ava went to Bologna; Ellery and Brigitte went to Copenhagen.

♦ Clubs and Careers

A diagram will help. See page 88.

1. A is not ABD *or* SP; B is not SL or PC; D is not SL or SP. (This information has been marked in.)

2. Consider C's remarks. He cannot himself be SL or SP. He knows that A is not SP (see #1). The only person who *can be* both SL and SP is E (see diagram). Therefore, C is saying that if he knew E was SL and SP, he would know that A was either Boomers or Longer Life.

He can only know this if he, C, is BL. Therefore, C must be BL. C is also saying that if he knew E was SP, he would know that B was either AC or PA. He can only know this if he, C, is GC. Therefore, C must be GC and BL must be GC.

		CLUBS				PROFESSIONS				
		Boom.	ABD	SL	BL	LL	AC	SP	PA	OM	GC
NAMES	A		×					×			
	B			×						×	
	C			×				×			
	D			×				×			
	E										
PROFESSIONS	AC										
	SP										
	PA										
	PC										
	GC										

3. For E to say that D must be Boomers or LL, he must know that D is not ABD. He can only know this if he, E, is ABD himself. Therefore, E is ABD. And for E to say that D must be OM or PA, he must know that D is not AC. Therefore, E must be AC. Therefore, ABD is AC.

4. Mark in also: PA is not SL and SP is not Boomers.

5. By elimination, we now know A is SL; B is SP.

6. We know B is SP, and we know B is either Boomers or LL. Therefore, either Boomers or LL is SP. But Boomers is *not* SP. Therefore, LL is SP.

7. By elimination, Boomers is PA and OM is SL. But A is SL. Therefore, A is OM and, therefore, by elimination, D is PA.

8. B is SP and SP is LL. Therefore, B is LL; therefore, by elimination, D is Boomers.

Complete result: A = SL = OM; B = L = SP; C = BL = GC; D = Boomers = PA; E = ABD = AC.

◆ The Five Discs

C reasons thus: If anyone were to see two red and two black, he would know that he was white. If anyone were to see two red, one black and one white, he would know that he could not be black, for, if he were, the man with the white disc would see two red and two black and would know that he was white. Similarly, if anyone were to see one red, two black and two white. If anyone were to see one red, one black and two white, he would know that he could not be black, for, if he were, either of the men wearing white would see one red, two black and one white, and would argue as above. If anyone were to see two red and two white, he would argue that he could not be black, for if he were, someone would see two red, one black and one white and would argue as above. If anyone were to see one red and three white, he would argue that he could not be black for if he were, someone would see one red, one black and two white and would argue as above; similarly, he would know that he could not be red. Therefore, if anyone sees me wearing red or black, he can deduce his color. Therefore, I must be white.

(This problem perhaps depends too much on assumptions about the relative intelligence of the people concerned. But some readers may find it an interesting piece of reasoning.)

◆ Gowns for the Gala

P, Q, R, S and T stand for the ladies; *l*, *m*, *n* and *o* stand for the materials of the gowns. /*l* means *not* lace, etc.

It will be convenient to put statements, predictions, etc., in shorthand form. (*Important:* Note that from "If X then Y," we can deduce "If /Y then /X," but we can*not* deduce "If /X then /Y.")

Q: If T*l*, then R*o*. If T*m*, then R*m*. Otherwise R*n*. Therefore, R/*l* under any conditions. (i)

R: R*l* unless S*m*. But we know R/*l*. Therefore, S wears *m*. (ii)

S: If P*l*, then S*o*. But we know S/*o*. Therefore, P/*l*. (iii)

P: P*l*, unless R*m* and S/*n*. If (R*m* and S/*n*), then P*o*. Therefore, if not (R*m* and S/*n*), then P*l*. (iv)

Therefore, P*l* or P*o*. But from (iii), we know P/*l*. Therefore, P wears *o*. (v)

And, from (iv), since P/*l*, therefore R*m* and S/*n*. Therefore, R wears *m*. (vi)

From Q, we see that the only conditions under which R wears *m* are if T wears *m*.

T says, if P/*l*, then Q*l* or Q*m*, whichever is "worn less by the rest of us." Therefore, Q wears *l*.

Complete solution: Priscilla, organdy; Quinta, lace; Rachel, muslin; Sybil, muslin; Tess, muslin.

◆ Salamanca Street

G, H, I, J and K stand for the people; g, h, i, j for the numbers of their houses.

Consider the fact that H's answers to questions are alternately Yes and No.

Answer to 1 and 2 cannot be No, Yes (<g> i), for we know that g< i; and the answers given are accepted by G and lead to a unique solution. Therefore, answers to the four questions must be Yes, No, Yes, No. (m3 means "a multiple of 3," /m3 means "not a multiple of 3," /sq means "not a sq.")

Therefore:

G thinks	I thinks	J thinks
1. > g	> g	< g
2. < i(49)	> i(49)	< i(49)
3. sq	sq	/sq
4. /m3	/m3	m3

From this information, G, I and J are able to arrive at unique answers. We now want to find what g must be to make this possible.

I must think 64, whatever g is (the only square between 49 and 99 that is /m3).

For G to arrive at a unique answer, which can only be 25, g must be between 24 and 16 inclusive. But we know that g is odd. Therefore, g must be 17, 19, 21 or 23.

J must think the answer is between 12 and g exclusive, and this answer must be m3 and not a square.

If g = 17, J thinks 15 (no alternative).

If g = 19, J thinks 15 or 18.

And if G = 21 or 23, there are more possibilities. As J arrives at a unique conclusion, therefore g must be 17 and J thinks h is 15.

K knows that only two of H's answers are true, and which, and he is able to deduce H's number correctly. K must think H's number is in one of these categories:

	I	II	III	IV	V	VI
1.	> 17	> 17	> 17	< 17	< 17	< 17
2.	< 49	> 49	> 49	> 49	< 49	< 49
3.	/sq	sq	/sq	sq	/sq	sq
4.	m3	m3	/m3	/m3	/m3	m3
Possibilities	many	81	many	imposs.	13, 14	none

The only category with a single possibility is II. Therefore, *H's number must be 81.*

Correct numbers: G, 17; H, 81.

Incorrect numbers of H's house: by G, 25; by I, 64; by D, 15.

◆ Around the Bend

(i) The possibilities, if the answer "Yes" is given to questions 1, 2 and 3, respectively, are:

1. 121, 144, 169, 196
2. 125
3. 116, 145, 174

Clearly, it is not possible for anyone to think that the answer is "Yes" to more than one question. But in order to say what he does, P must think the answer is "Yes" to one of them. It cannot be Question #2, for P would then claim to know V's number with certainty.

(ii) The possibilities are:

either V has answered "Yes" to Question #3, so that if P knows the number is greater than 150 he will know it is 174;

or V has answered "Yes" to Question #1, and P's number is either 169 or 196 and, if P knows that the number is greater than 150, he will know V's number is the other.

(iii) I knows the questions and answers, and has heard P's comment; he also, of course, knows his own number. If V has answered "Yes" to Question #3, I will have no information about P's number (except that he has reason to believe that the difference between his number and P's is less than 30) and could not claim to know it. Therefore, the answer must have been "Yes" to Question #1, and not to Question #3. And I knows that P must be 169 or 196 (in order to say what he does). And I thinks that V is 121 or 144. (The answer to Question #1 must have been "Yes," and I has reason to believe that V is less than 150.)

(iv) With this information, his own number, and the belief that the difference between his and P's is less than 30, I claims to know the numbers of both the other houses. He can only do this if his number is 144, and he will then say that V's is 121, and that P's is 169.

Therefore, I's number is 144, and since I is correct about P's number, therefore P's number is 169.

(v) Since only V's answer to Question #2 was true, therefore V's number is not a square but is m29; and since I's belief that V's number was less than 150 is incorrect, therefore V's number must be 174.

Therefore, numbers are: Proper, 169
Improper, 144
Vulgar, 174

◆ The Island of Indecision

1. Consider B(ii). If true, all E's statements are false. Therefore, A is a Q (E[ii]), and today is Tuesday; therefore, A's statements are true. Therefore, B is an O and today is the 14th. Therefore, B's statements are all false. But this is contrary to our original assumption (B[ii] true). Therefore, assumption wrong. Therefore, B(ii) false. Therefore, E is not N.

2. Consider B(iii). If true, the date is odd, and in that case, all B's remarks would have to be true. But we know that B(ii) is false. Therefore, B(iii) is false. Therefore, B is not O. Therefore, B(i) is false (there are no conditions under which one true remark is followed by two false). Since B(iii) is false, therefore A(ii) is false.

3. B's remarks are all false. Therefore, B is not Q (see conditions); therefore, D(i) is true. Therefore, D's remarks are either all true or alternately true and false. Therefore, D(iii) is true. Therefore, C is P.

4. Therefore, C(i) is false (Ps only tell the truth on Wednesdays and Fridays). And since Ps' remarks are either all true or all false, therefore they are all false. Therefore, E is not P; B is not P. And since Ps tell the truth on Wednesdays and Fridays and C's remarks are all false, therefore today is not Wednesday, Friday or Monday [C(i)].

5. Therefore D(iv) is false. Therefore, D's remarks are alternately true and false. Therefore, D is Q and the day must be Monday or Thursday. But it is not Monday, so it must be Thursday.

6. Therefore, E(i) is true. Therefore, E(iii) is true. Therefore, A is Q. Therefore, E(ii) is false. Therefore, E is Q (since statements are alternately true and false).

7. Since A is Q, today is Thursday and A(ii) is false. Therefore, A(i) is true. Therefore, the date is the 14th.

8. B is not P [C(iii) false], and not O [B(iii) false], and not Q (remarks by a Q cannot all be false); therefore, by elimination, B is N.

Complete solution: Adolf is Q; Basil is N; Clyde is P; Dean is Q; Elmer is Q.

The day is Thursday, the 14th.

◆ Country Crescent

Cuthbert is told two things:

(i) the numbers of "their three houses" add up to twice the number of his house;

(ii) the product of the three numbers is 1260.

But this doesn't tell him what the numbers are. Therefore, there is more than one possibility.

Some trial of the possible sets of three factors of 1260 is necessary here. We want to find different sets whose sum is the same, and this sum must be even (the numbers add up to twice the number of his house).

Remembering that there is no number 1, it will be found that the only two such sets are 4, 9, 35 and 5, 7, 36 (sum 48 in both cases).

These two possibilities must therefore be in Cuthbert's mind when he makes his remark.

Casey's next remark then tells Cuthbert two things:

(i) the number of his (Casey's) house is greater than that of anybody else's;

(ii) that this information will enable Cuthbert to choose between the two alternatives.

Therefore, Casey's number must be 36; the numbers of "those three" must be 4, 9, 35; Cuthbert's number must be 24 [(4 + 9 + 35) ÷ 2].

◆ Index

Add and Subtract #1 and 2, 19, 20
Addition and Division, 22
Ages, 10, 24–25, 27
Ambitions, 26–27
Architect's Shirt, 8
Around the Bend, 44
Birthdays, 37
Christmas Compensation Club, 24
Clubs and Careers, 40–41
Competition, 35
Country Crescent, 46
Easter Parade, 38–39
Family relations, 32
Five Discs, 41–42
Gowns for the Gala, 42
Hats, 38–39
Hitler and Goering, 14
House numbers, 10, 25, 27–28, 43, 44, 46
Intellectual Awareness, 35
Intelligence Test, 36
Island of Imperfection, 31
Island of Indecision, 45
Long Division, 14–19, 21–22
Jobs, 32
Lies, 31, 38–39

Marriages, 8, 31–32
Math puzzles, 13–22
Monogramy Comes to the Island, 31–32
Murder mysteries, 9, 12
Out the Window and Over the Wall, 28–29
Predictions, 8
Prison break, 28–29
Professions, 11, 40–41
River Road, 10
Salamanca Street, 43
Schoolmates, 11
Shirts, 8
Sinister Street, 25
Test, intelligence, 36
Tom, Dick and Harry, 31
Tribes, 30, 32
Trips Abroad, 39–40
Uncle Knows Best, 27–28
Uncles and Cousins, 34
Vacations, 39–40
Vests and Vocations, 11
Who Killed Popoff?, 9
Willahs and the Wallahs, 30
Years Roll By, 24–25
Youthful Ambitions, 26–27